Fabricio Araujo

Simulating data from mapping by sequencing experiments

Fabricio Araujo

Simulating data from mapping by sequencing experiments

A pipeline for simulating data from mapping by sequencing experiments in Arabidopsis thaliana

ScienciaScripts

Imprint
Any brand names and product names mentioned in this book are subject to trademark, brand or patent protection and are trademarks or registered trademarks of their respective holders. The use of brand names, product names, common names, trade names, product descriptions etc. even without a particular marking in this work is in no way to be construed to mean that such names may be regarded as unrestricted in respect of trademark and brand protection legislation and could thus be used by anyone.

Cover image: www.ingimage.com

This book is a translation from the original published under ISBN 978-3-330-77113-0.

Publisher:
Sciencia Scripts
is a trademark of
Dodo Books Indian Ocean Ltd. and OmniScriptum S.R.L publishing group

120 High Road, East Finchley, London, N2 9ED, United Kingdom
Str. Armeneasca 28/1, office 1, Chisinau MD-2012, Republic of Moldova, Europe
Managing Directors: Ieva Konstantinova, Victoria Ursu
info@omniscriptum.com

Printed at: see last page
ISBN: 978-620-8-63565-7

SUMMARY

THANKS

To Professor Dr. Sylvain Henri Darnet for his valuable guidance and wisdom.

To friends and family for their support.

To the members of the Plant Biotechnology Laboratory for the information exchanged.

To CNPQ for the grant.

To the Graduate Program in Genetics and Molecular Biology for their professionalism and promptness.

To the Federal University of Para for the infrastructure provided.

SUMMARY

Recent advances in next-generation sequencing (NGS) technologies have reduced the time needed to identify mutations in *forward* genetics, one of the main ways of characterizing gene function in plants. The methodology used to identify mutations induced by physical or chemical mutagens in a *mapping* population is known as *mapping-by-sequencing*. Among the various techniques that use this methodology, *Next-generation mapping* (NGM) stands out because it requires a small number of mutants in the mapping population. However, in the published works, it is not clear why the parameters used to form the mapping population (number of individuals, number of mutants, way of crossing) and for sequencing (platform used, sequencing method, coverage) were used in the experiments. For this reason, a bioinformatics *pipeline* was developed that enables the simulation of data from experiments using the NGM technique with the model plant *Arabidopsis thaliana*, one of the most studied and sequenced plants in the literature. The pipeline developed, NGMdatasim, encompasses all the phases of an NGM experiment and allows the customization of parameters to improve the design of new experiments. With this, it was possible, for example, to identify that reducing the mapping population of an experiment did not interfere with the final result, which could make it possible to reduce costs in new experiments.

Keywords: *forward* genetics, next generation sequencing, mapping population, *Arabidopsis thaliana*, data simulation.

1 INTRODUCTION

Rapid progress in the development of genetic tools has led to the creation of powerful approaches for mapping between observable traits of interest (phenotype) and the genetic construct (genotype) causing these traits (Leshchiner *et al*, 2012). Specifically in plants, the application of these approaches to genetics and plant breeding is becoming increasingly popular (Rafalski, 2010). In *Arabidopsis thaliana*, for example, these traits of interest can be related to changes in plant morphology, changes in growth conditions, resistance or susceptibility to chemical agents, etc. (Meinke, 2013).

These approaches to the systematic association between phenotype and genotype can be grouped into *forward* and *reverse* genetics. In simplified terms, *forward* genetics deals first with the selection of a phenotype of interest (mutant) and then with the identification of the genetic region that contributes to the phenotype. *Reverse* genetics, on the other hand, deals first with the analysis of a gene of interest and its possible mutations and then looks at the phenotypes produced (Jander *et al*, 2002). Although *reverse* genetics has become very popular in recent decades, it lacks functional resources (focus on phenotypes of interest). In addition, unlike *reverse* genetics, *forward* genetics makes it possible to identify a broad spectrum of mutant alleles (Austin *et* al, 2011).

The genetic variability (mutations) of organisms, responsible for phenotypes, occurs spontaneously and slowly in nature. Therefore, the use of mutagenic agents, whether physical or chemical, can accelerate this process. The induction of genetic variability through the use of mutagens is commonly used in *forward* genetics studies (Page & Grossniklaus, 2002). The chemical mutagen ethyl methanesulfonate (EMS) has been used on a large scale to induce mutations and possibly discover the functionality of many genes (Hartwig *et al*, 2012). EMS is an alkylating agent that tends to induce point mutations in the entire genome, called SNPs (*Single Nucleotide Polymorphisms*) (Lindner *et al*, 2012). About 99% of the time, EMS induces C>T and G>A substitutes and less frequently, G>C and C>G or G>T and C>A substitutes (McCallum *et al*, 2000).

Although the induction of mutations using mutagenic agents can be applied to various organisms, this type of study is most commonly used in model organisms due to the ease of inducing mutations, the availability of a large number of genetically well-defined strains and the availability of genome manipulation tools for testing and characterizing candidate genes and mutations (Schneeberger, 2014). In addition to these aforementioned characteristics, a short life cycle, self-compatibility and easy growth in greenhouses make *Arabidopsis thaliana*, a member of the *Brassicaceae* family, a widely accepted model species and a major object of study in plant research (Verbruggen, Hanikenne & Clemens, 2013).

Arabidopsis thaliana has a genome of approximately 135 Mb, 27,416 coding genes, 1,359 non-coding genes and 924 pseudogenes. Its genetic structure has a functional diversity similar to that of *Drosophila* and *Carnorhabditis elegans*, other model eukaryotic organisms. This was the first plant genome to be completely sequenced and provided a basis for more comprehensive comparisons between conserved eukaryotic processes, as well as the discovery of various gene functions in plants (Arabidopsis Genome Initiative, 2000). Over time, the *Arabidopsis thaliana* genome has become one of the most sequenced and best curated, with genes being well annotated and characterized (Koornneef & Meinke, 2010).

The discovery of SNPs can be done reliably with the use of next-generation sequencing technologies through whole genome sequencing. It should then be possible to directly identify the induced variants for a mutant phenotype. However, the induction of mutations by EMS generates several mutations per Kb (one mutation every 112 bp to one mutation every 171 Kb) (James *et al*, 2013), so that direct sequencing of the mutant genome is not enough to discover the mutation causing the phenotype studied (Schneeberger & Weigel, 2011). It is in this context that a new approach to *forward* genetics has been developed: *mapping-by-sequencing*.

1.1 IDENTIFICATION OF EMS MUTATIONS IN *FORWARD* GENETICS : *MAPPING-BY-SEQUENCING*

As seen above, *forward* genetics is a widely used approach for discovering gene function in plants and other organisms. This tool can be used in various types of experiments and can be used separately or as part of a protocol or methodology (James *et al*, 2013).

The methodology that uses genetic screening where mutants, generated by *forward* genetics, are isolated based on their phenotypes and use *bulked segregant analysis* (BSA) is called *mapping-by-sequencing* (Schneeberger *et al*, 2009). BSA is a method for the rapid identification of markers in specific regions of the genome. The method involves comparing two *pooled* DNA samples from individuals in segregating populations originating from a single cross. In each *pool*, or *bulk*, the individuals are identical for the phenotype or gene of interest but are arbitrary for other genes. Two different *pools* for the same trait (for example, resistant or susceptible to a particular disease) are analyzed to identify markers that differentiate them (Michelmore, Param & Kesseli, 1991).

In order to use the BSA method, it is necessary to create a segregated mapping population containing the phenotype of interest. The creation of this population is directly linked to the concept of the experiment used by *mapping-by-sequencing* techniques, since they are related to the time and cost of the experiment. Figure 1 shows three different scenarios for forming the mapping population.

1.1.1 Direct sequencing of individual mutant genomes

In the direct sequencing of individual genomes of mutants, a mapping population is not formed. The sequencing of individuals is carried out individually on the mutants (M) of a population originated by classical crossing or, more commonly, backcrossing (Figure 1-III) (Schneeberger, 2014).

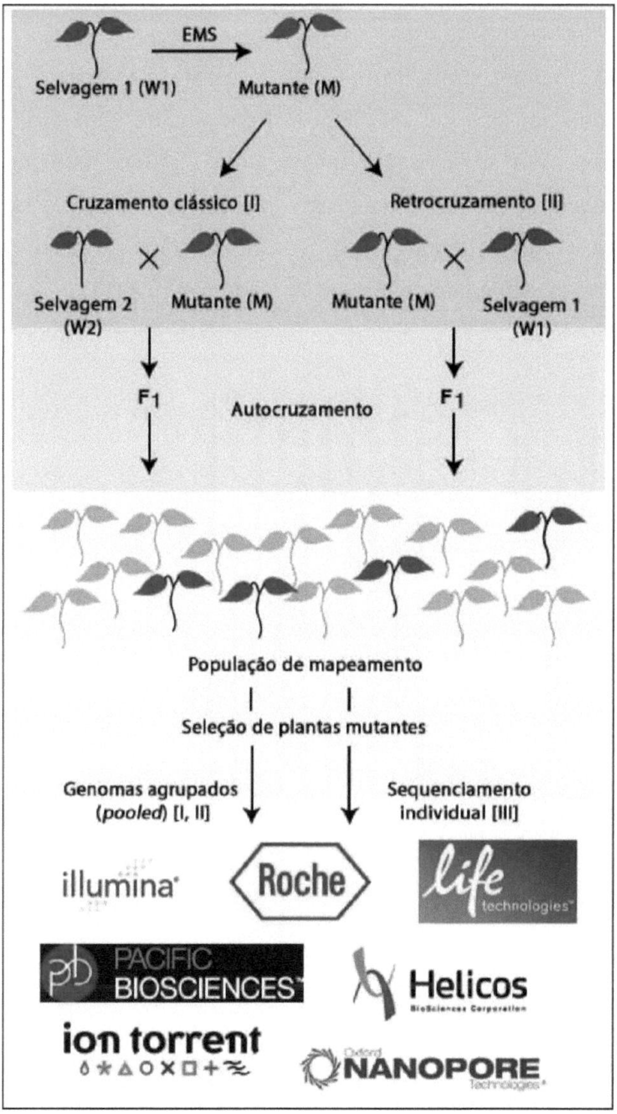

Figure 1. Overview of the different strategies for forming the mapping population.

I - Formation of the mapping population by classical crossing or *outcrossing*.

II - Formation of the mapping population by isogenic crossing or *backcrossing*.

III - Direct sequencing of mutants. Adapted from (James *et al*, 2013)

1.1.2 Classical population mapping or *outcrossing*

In classic mapping, an EMS mutant strain (M) is used to create a mapping F2 population by *outcrossing* between this mutant strain (M) and another genetically different strain (W2) followed by a *self-crossing* shift Figure 1-I) (Schneeberger, 2014). In this type of population, the mutant plants carry the mutation of one of the parents. Sequencing the *pool of* mutant genomes will reveal the bias in the frequency of alleles that are linked to the causal mutation (Figure 2) (James *etal*, 2013).

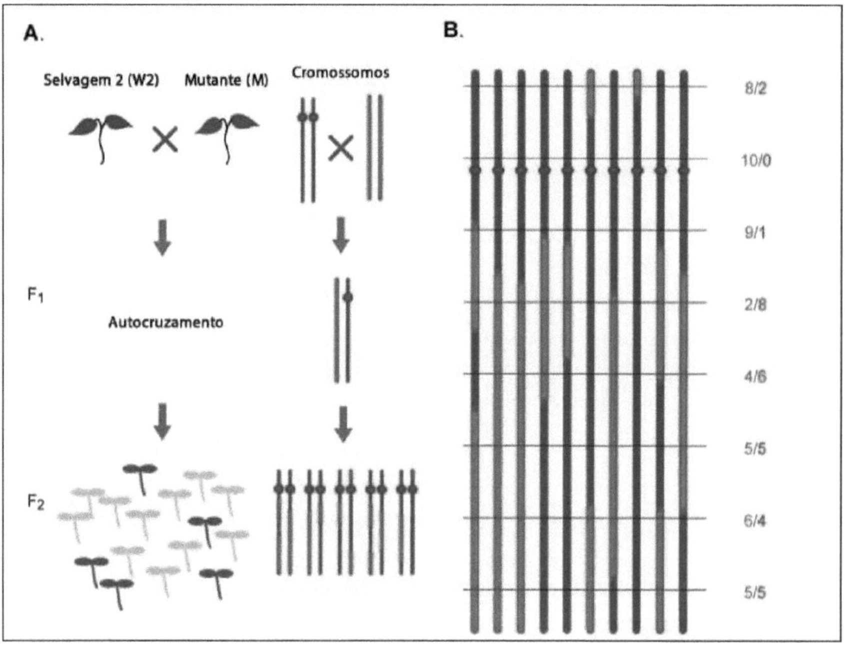

Figure 2: Genetics of classic crossbreeding.

A - Generation of the mapping population by *outcrossing*. The mutant plants carry the mutation of one of the parents. B - Sequencing the *pool of* mutant genomes will reveal the bias in the frequency of alleles that are linked to the causal mutation. Adapted from (Schneeberger & Weigel, 2011)

1.1.3 Isogenic population mapping or *backcrossing*

In isogenic mapping, a cross is made between mutants (M) and their non-mutant parents (W1). The F2 offspring derived from this cross has a minimum number of the segregating *loci* responsible for the phenotypic change (Schneeberger, 2014). In this type of population, any mutation that is not linked to the chosen phenotype will segregate at a rate of 1:3 in the population while mutations linked to the chosen phenotype will segregate at a rate of 1:1 in the population (Figure 3) (Lindner *et al*, 2012).

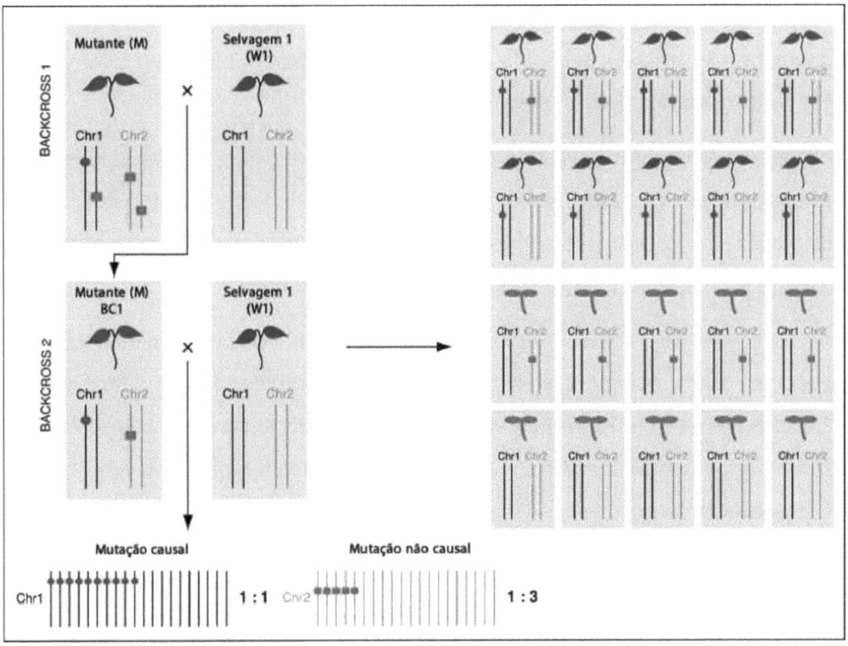

Figure 3: Genetics of isogenic crossing.

A mutant plant is crossed with its non-mutant parent. In the first *backcrossing*, around 50% of the SNPs not related to the mutation will be eliminated. After the second *backcrossing*, the mutations unrelated to the phenotype of interest will segregate at a rate of 1:3 in the population while the causal mutation will segregate at a rate of 1:1 in the population. Adapted from (Lindner et al, 2012)

Recently, new studies have placed more emphasis on the creation of isogenic populations. Abe and colleagues (2012) demonstrated the ability of this type of mapping to identify agronomically important traits in rice that could not be identified in a classic mapping population, as there are subtle traits that can be masked as natural genetic variation.

1.2 MAPPING-BY-SEQUENCING IN THE LITERATURE

After forming the mapping population and sequencing the *pool* of selected mutants, it is necessary to analyze the data generated. Several tools and analysis *pipelines* have been developed and applied to various model species, including plants, yeast, nematodes, mammals and invertebrates for the detection of mutant genes in *forward* genetics. When it comes to specific tools, Table 1 shows a representative history of tools developed to analyze the data generated by experiments using DNA-based *mapping-by-sequencing* methodology.

The first tool developed for analyzing *mapping-by-sequencing* data was SHOREMap (Schonneberger *et al*, 2009), which identified a mutation in *Arabidopsis thaliana* responsible for slow growth and light-green leaves in an unknown gene. The tool was used to sequence a *bulk of* a mapping population of 500 individuals.

In 2011, Austin and colleagues developed a new approach to analyzing the *mapping-by-sequencing* methodology. Called *Next-generation Mapping* (NGM) (Austin *et al*, 2011), this tool was able to identify three genes related to the cell wall biology of *Arabidopsis thaliana* with a *bulk* from a sequencing population of 50 individuals. This tool will be detailed later.

Table 1. Representative history of tools developed using DNA-based *mapping-by-sequencing* methodology.

Tool	**Year**	**Species for which it was developed**
SHOREMap (Schneeberger *et al*, 2009)	2009	*Arabidopsis thaliana*
Next-generation mapping (NGM) (Austin *etal*, 2011)	2011	*Arabidopsis thaliana*
CloudMap (Minevich *etal*, 2012)	2012	*Caenorhabditis elegans and Arabidopsis thaliana*
Fishyskeleton (Bowen *etal*, 2012)	2012	*Danio rerio*
MegaMapper (Obholzer *et al*, 2012)	2012	*Danio rerio*
Multipool (Edwards & Gifford, 2012)	2012	*Saccharomyces cerevisiae*

MutMap (Abe *etal*, 2012)	2012	*Oryza sativa*
MutMap+ (Fekih *etal*, 2013)	2013	*Oryza sativa*

Source: Adapted from (Schneeberger, 2014), revised and updated on 01/02/2016.

Minevich and colleagues developed the CloudMap tool (Minevich *et al*, 2012) which was able to identify a mutation in the *vab-3* gene of *Caenorhabditis elegans* in a *bulk of* a mapping population of 50 individuals. In the same work, the same tests carried out in (Schonneberger *et al*, 2009) were reprocessed using the tool and were also successful.

Bowen and colleagues developed a database of SNPs from a mapping population of 20 *Danio rerio* individuals with five uncharacterized mutations (*minamoto, welded, hollow, lruehrentner and Schrumpfkopf)* (Bowen *etal*, 2012).

Obholzer and colleagues developed the MegaMapper tool (Obholzer *et al*, 2012) which was able to identify mutations in the *lmx1b* and *jagged1b* genes of *Danio rerio* that cause abnormal ear morphogenesis, in a *bulk of* a mapping population of 100-200 individuals.

Edwards & Gifford developed the Multipool tool (Edwards & Gifford, 2012) which was tested on mutations already characterized in yeast. Mutations related to resistance to the chemical agent 4-NQO and heat tolerance were identified by the tool in a *bulk of* a mapping population of approximately 10000 individuals.

Abe and colleagues developed the MutMap tool (Abe *et al*, 2012) which was able to identify single genomic regions that are most likely to carry the mutations related to light green leaves and seminanism, an agronomically important trait, in *Oryza sativa*. The MutMap experiments were carried out on a *bulk* mapping population of 20 individuals. An extension of this tool, called MutMap+ (Fekih *et al*, 2013) was developed by Fekih and colleagues. This tool was able to identify a region responsible for the chromosome 1 mutation in the *Hit9188* mutant of *Oryza sativa*, characterized by dwarfism, light green leaves and premature death. This work was carried out using a *bulk* mapping population of 40 individuals.

Of the tools listed above, only NGM and SHOREMap have been applied in other studies, which is why they are so important in the context of this work. Hartwig and colleagues, for example, were also able to identify mutations in the *lhp1* gene of *Arabidopsis thaliana* by sequencing a *bulk of* a population of 270 individuals on the *Ion Torrent* platform using SHOREMap (Hartwig *et al*, 2012). In another example, using the HiSeq2000 sequencing platform, a bulk population of 26 individuals and the NGM tool, Golas and colleagues identified five mutations in the potato rpi-dlc2 gene related to resistance to the pathogen *Phytophthora infestans* (Golas *et al*, 2012).

When analyzing the tools and *pipelines* described above, it is not clear why a particular sequencing technology is used over another, neither the sequencing parameters used nor the number of mutants used to form the mapping population. This may be why there are several tools for analyzing mutant genomes but so few studies using these tools to discover other new mutations.

1.3SEQ∪ NEW GENERATION START-UP

The first generation of DNA sequencing is known as the Chain Ending Method developed by Sanger (Sanger, Nicklen & Coulson, 1977). This method dominated the industry for almost two decades of the last century and led to major advances, including the completion of the first sequence of the human genome (International Human Genome Consortium, 2004). Although Sanger sequencing is highly accurate, it is expensive and time-consuming, and very labor-intensive in large sequencing projects. In this sense, efforts have been directed towards the development of new technologies. These new sequencing technologies are known as *Next-generation sequencing* (NGS) ($Metzker_1$ 2010). NGS methods are high-throughput DNA sequencing technologies that allow the sequencing of millions of DNA strands in parallel, generating large volumes of data in a short period of time and at a low cost per bp (Pettersson, Lundeberg & Ahmadian, 2009).

The arrival of these new technologies on the market has changed the way we think about scientific approaches in basic, applied and clinical research (Metzker, 2010). The biggest advance provided by NGS is the production of a massive amount of data at a lower cost compared to Sanger sequencing. Figure 4 shows the evolution of the cost of sequencing human genomes. The decrease in the cost of sequencing has a direct implication on the creation of more sequencing projects, as shown in Figure 5.

The possibility of sequencing entire genomes of organisms allows comparative and evolutionary studies to be carried out on a large scale in a way that was unimaginable a few years ago. The variety of characteristics of NGS technologies allows several platforms to coexist on the market, with some having advantages over others in particular cases (Metzker, 2010). This paper will look at four platforms: pyrosequencing (Roche/454), sequencing *by oligo* ligation *detection (*SOLiD), sequencing by reversible terminators (Illumina) and sequencing by ion semiconductor (Ion Torrent). These platforms were chosen because of their representative history in sequencing projects since the 1990s.

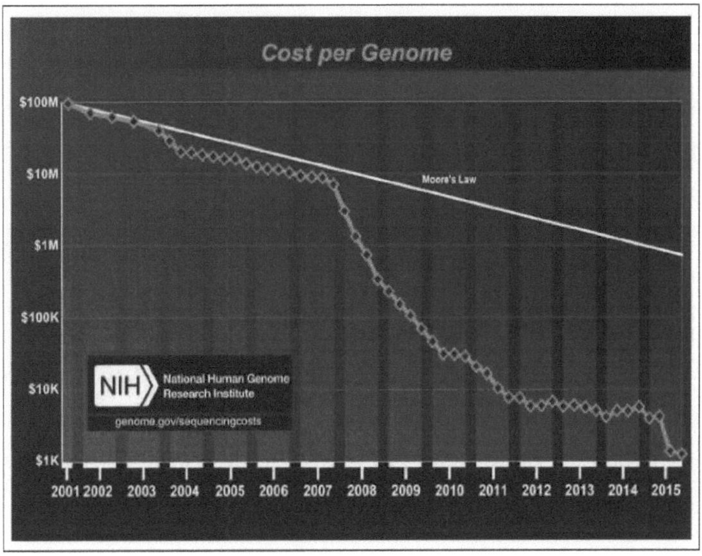

Figura 4. Cost of sequencing human genomes over the years.

The average cost of sequencing the human genome in 2001 was $95,263,072 (the cost per Mb being $5,292.39). After 15 years of advances in NGS technologies, this figure drops to an average of $1,245 (the cost per Mb being $0.014). Adapted from http://www.genome.gov/sequencingcosts/

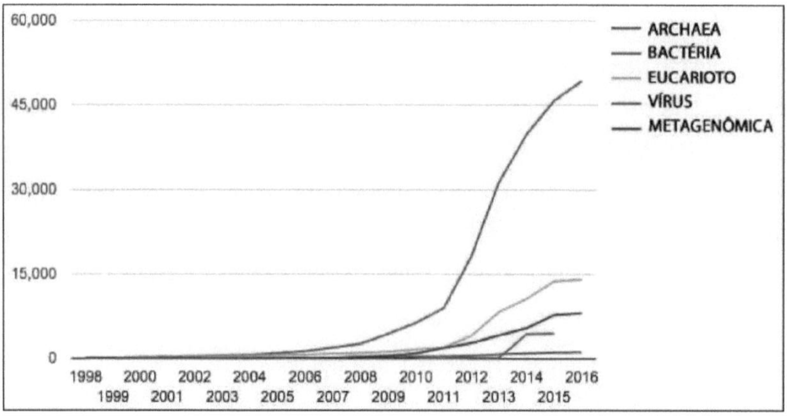

Figura 5. Number of sequencing projects over the years.

In 1998 there were approximately 55 projects mainly related to bacteria. After 17 years of technological advances in NGS and decreasing costs of sequencing, the current number of projects is around 80,000, related to *archaea*, bacteria, eukaryotes, viruses and metagenomics. Adapted from http://sulab.org/2013/06/sequenced-genomes-per-year/

The choice of sequencing platform for an experiment can be subject to various factors. Therefore, a comparative analysis of the platforms in terms of sequencing methodology, run time, number of reads per run (in millions), number of bases per read, amount of data generated (in Mb) per run, reagent cost per run and reagent cost per Mb is shown in Table 2.

These platforms concentrate their efforts on generating a larger number of short reads. Longer reads (>500 bp per read) are better because longer reads are more easily assembled than shorter reads (100 to 300 bp per read). However, the work and costs of technologies that generate longer reads are higher.

Table 2. Comparative analysis of new generation sequencing platforms.

Platform	Methodology Sequencing	Race time	Millions of reads/run	Bases/read	Mb/race	Reagent cost / run	Cost of reagent/Mb
Roche 454 Junior	Summary (Pyrosequencing)	08 h	1	500	900	$1400	$7
SOLID 5500x1	Link	8 days	>700	110	77000	$6101	<$0.08
Illumina HiSeq 2000	Summary	2 days	320	100	96000	$1362	$0.12
Ion Torrent 318 chip	Synthesis (detection of H)$^{+}$	2 hours	4-8	>100	>1000	$925	$0.93

Source: Adapted from (Glenn, 2011), revised and updated on 01/09/2015.

1.3.1 Sequencing errors

None of the technologies mentioned above has been able to generate 100% accurate readings on each basis. The most common errors found in NGS technologies are:

- Substitution errors: occur when a particular nucleotide in a DNA sequence is replaced by another nucleotide in the same position. Substitution errors can be transitional (A/G to C/T) (Guo *et al*, 2012) or transversional (A/C to A/T, G/C to G/T) (Brodin *etal*, 2013).
- Insertion and deletion errors (*indel*): insertion errors occur when a nucleotide is incorporated into a sequence that is not part of the original chain, increasing the overall size of the chain by at least 1 bp. Deletion errors are the opposite of insertion errors; they occur when a nucleotide is removed from a chain, decreasing the overall size of the sequence by at least 1 bp (Bragg *etal*, 2013).
- Homopolymer errors: these are errors of interpretation. Many DNA sequences have regions where the same base is repeated several times in succession (AAAA, GGGG, etc.). These regions are known as homopolymeric regions. Thus, sequencing these regions generates a very high light or electrical signal that can be interpreted as a single base or a homopolymer, depending on the cut-off value set for the sequencing (Bragg *et al*, 2013).
- The effects of sequencing errors can be reduced by sequencing the genome

repeatedly, achieving greater *coverage*, which refers to the average number of times each base is sequenced. In addition, error rates tend to be higher at the end of the NGS technologies' lifetimes (Schirmer *et al.*, 2013).

1.3.2 BiasGC

Genomes are commonly related to their GC content (proportion of the genome made up of guanine and cytosine). It is not common for the content of A, T, G and C to be similar; on the contrary, some organisms have a global GC content ranging from 16% to 75% (Lightfield, Fram & Ely, 2011). In experiments, there is usually a bias related to high or low GC content that leads to disproportionate coverage of these regions, which requires high sequencing coverage to obtain a minimum coverage value that avoids errors and confusions in the analysis of the data generated (Dabney & Meyer, 2012). This sequencing bias, or GC bias, can be verified at any stage of the sequencing experiment. In addition, GC bias can vary between laboratories, sequencing experiments and even between lanes of the same flow cell. Variables related to the amplification stage, such as the specific type of DNA polymerase used and the cycle temperature, are pointed out as the main players in GC bias, although the actual mechanism for its occurrence is not fully known (Lightfield, Fram & Ely, 2011).

1.4 DATA SIMULATION

Evaluating different types of *mapping-by-sequencing* experiments requires the development and sequencing of thousands of mapping populations, which is not practically possible. However, *in-silico* simulations make it possible to generate many experiments for testing, evaluating algorithms, optimizing parameters and possibly generating better experiment designs (James, 2013).

Before starting a *mapping-by-sequencing* experiment, it is necessary to evaluate various parameters, such as: the size of the mapping population, the concentration of EMS to be used (to generate more or fewer mutations), the sequencing coverage required, the size of the sequencing reads needed, the best sequencing platform for the experiment, the use of *paired-end* or *single-end* reads in the sequencing, etc. These and other characteristics can be specified, in part, by simulations.

In order to generate reliable simulated data, it is not enough just to generate random sequences of a reference genome or to generate random crosses to generate a population, since there are a number of variables related to each stage of an experiment. It is therefore necessary to create more sophisticated ways of generating reliable simulator data (Sun, Peng & Mumm, 2011).

2. OBJECTIVES

2.1 GENERAL

To develop a simulation *pipeline* NGM experiments on *Arabidopsis thaliana* (simulation of the mapping population, application of mutations and NGS runs) that will enable testing, evaluation of algorithms, optimization of parameters and possibly generate better experimental designs.

2.2 SPECiFIC

- Define and implement the *pipeline* steps*:* simulation of the mapping population, simulation of mutants in the mapping population, NGS runs of the mutant genomes and mapping between the NGS reads and the *Arabidopsis thaliana* reference genome;

- Generate the specific file for processing in the NGM tool;

- Develop a graphical interface for the *pipeline* that can be accessed via the Internet;

Validate the *pipeline* by comparing the simulated data generated with the real data from the work of (Austin *et al*, 2011);

3. MATERIALS AND METHODS

The *pipeline* developed in this work, called NGMdatasim[1], consists of an integration of tools. In it, all the steps of an NGM experiment are implemented: simulation of the mapping population and mutants (Figure 6-1 and 6-2), seen in section 1.4.1, simulation of the NGS run (Figure 6-3), seen in section 1.4.2, mapping with the reference genome (Figure 6-4), file conversions (Figure 6-5) and generation of the NGM-specific file (Figure 6-6), which will be described in this section.

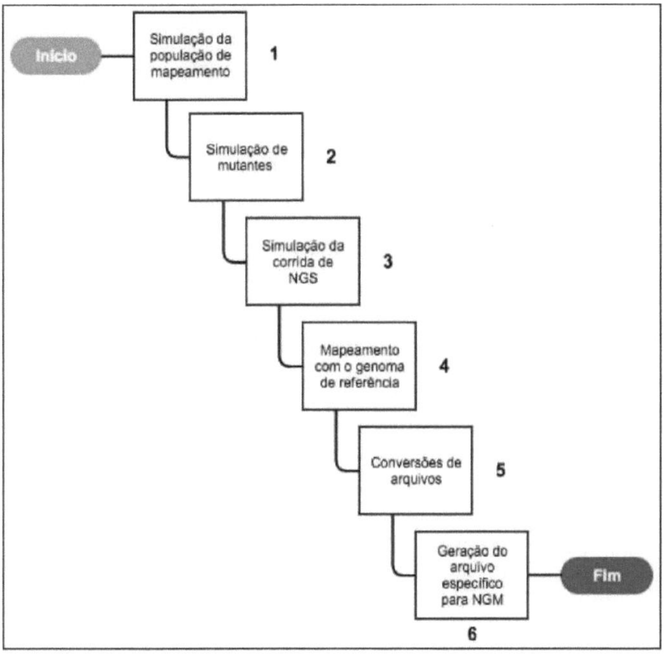

Figure 6. NGMdatasim *pipeline.*

Description of the pipeline steps; 1,2- Simulation of the mapping population and mutants using the Pop-seq Simulator tool; 3 - Simulation of the NGS run using the ART and BEAR simulators; 4 - Mapping with the reference genome using the Bowtie tool; 5 - File conversions using the SAMtools tool; 6 - Generation of the specific file for NGM using the SAM2NGM.pl script.

3.1 DATA SIMULATION

3.1.1 Mapping population simulation

In the work by (James *et al*, 2013), a *pipeline* is developed for simulating *mapping-by-sequencing* experiments in *Arabidopsis thaliana*. This *pipeline* is called Pop-seq Simulator.

Initial parental genomes are generated using a number of monozygotic markers specified by the user. These markers are placed randomly or at a specific *locus*. If the marker under selection is not specified, then one of the other markers is randomly selected as the causal mutation and is used for selection at the end of each mutant. In the same way, a wild genome is simulated with the exception of the *loci* of the marker with the wild allele. To simulate offspring genomes, recombined haploid genomes are combined from one or two virtual parents. The offspring genomes are used as parents in the next crosses. During each cross, virtual gametes are generated from each genome by determining the number and location of the recombination involved. The actual number of recombinations per meiosis for each chromosome is randomized based on the distribution of recombination events in *Arabidopsis thaliana* (James *et al*, 2013).

This simulator has been developed in the Perl programming language, is in its 1.0 version and is available for free *download*. It allows various parameters to be modified, so that the simulation can be very customized. The parameters accepted by the simulator are:

n: number of plants in the population;

- a: number of mutant plants to be generated (use n or a);
- f: type of crossing;
- m: file with mutation markers (tabular text file where the first column must be an identifier, the second column the marker chromosome, the third column the position of the mutation, the fourth column the reference base and the fifth column the mutant base);

There are also other optional parameters that can be used (Figure 7). However, the use of these extra parameters was not necessary when conducting the experiments in this work. For this work, the Pop-seq Simulator will be used to simulate the recombinant genomes of the mapping population.

```
usage:perl pop_seq.pl -n 10 -f F2 -m Input_data/Ler-1.5000.txt -o genotyping.txt

Required:
 -n Number of plants in the population
 -a Number of mutant plants to generate (Either -n or -a)
 -f Crossing scheme in "litral words" superated by ":" . For example: F2 or B2 or F2:B1:F1
(This make BC1F2 by one selfing F2 after BC1). If the last cross is BC then it will only pr
oduce 3 plants
 -m marker file

Optional:
 -h help
 -o output file (default is genotyping.txt)
 -l print whole population from every generation
 -s mutate site (example: 1:100 for chr1 and 100bp. Default is: 1:100000)
 -i genotype file from previous simulation
 -e phenotyping error in percentage (based on total number of plants produced in a populati
on)
 -c config file for species than arabidopsis (Default is Arabidopsis)
 -d Use single selfed(F) genome till the last genration and create last population from it
 -j Use single backcrossed(B) genome, default is 3
 -V verbose
```

Figure 7. Using the Pop-seq *Simulator.*

The mandatory parameters are used to generate the number of plants in the population (-a) or the number of mutant plants (-n), identify the type of cross to be used (-f) and indicate the file containing the markers of the mutations that will be applied (-m).

3.1.2 Simulation of NGS runs

Once the mapping population data has been generated, sequencing on an NGS platform is required. In the same way that the generation of the mapping population is done by simulation, the NGS runs will also be done. Two simulators will be used to simulate the runs: ART Simulator (Huang *et al*, 2011) and BEAR Simulator (Johnson *et al*, 2014).

ART Simulator is a set of simulation tools, developed in C++, which generates synthetic NGS sequencing reads. The simulation is done by emulating the sequencing process with technology-specific read error models and empirically parameterized base quality value profiles summarized from a large amount of recalibrated sequencing data (Huang *et al*, 2011). The ART simulator generates data from the 454, Illumina and SOLiD sequencing platforms.

For each platform there is a different tool: for 454 there is the art_454 tool (version 2.6.0), for SOLiD there is the art_SOLiD tool (version 1.3.3) and for Illumina there is the art_illumina tool (version 1.3.3). Each tool has a different set of parameters relating to the technology for which it was developed.

The art_454 tool allows the simulation of *single-end* runs, *paired-end* runs and *amplicon* sequencing. Parameters are required (Figure 8):

- INPUT_SEQ_FILE: file in FASTA format containing the DNA or RNA reference sequences;
- OUTPUT_FILE_PREFIX: the prefix of the files generated by the simulator;
- FOLD_COVERAGE: the sequencing coverage of the reference genome;
- MEAN_FRAG_LEN: o average DNA fragment size for *paired-end* race simulations ;
- STD_DEV: o standard deviation for o DNA fragment size for *paired-end* run simulations ;
- #READS_PER_AMPLICON: number of *reads* per *amplicon* ;
- #READ_PAIRS_PER_AMPLICON: number of *read* pairs per *amplicon* ;

```
====================================================================
                    ART_454 (Version 2.6.0)
                 Simulation of 454 Pyrosequencing
     Copyright (c) 2008-2015, Weichun Huang. All Rights Reserved.
====================================================================

===== USAGE =====

SINGLE-END SIMULATION
        art_454 [-s] [-a ] [-t] [-r rand_seed] [ -p read_profile ] [ -c num_flow_cycles ] <INPUT_SEQ_FILE>
<OUTPUT_FILE_PREFIX> <FOLD_COVERAGE>

PAIRED-END SIMULATION
        art_454 [-s] [-a ] [-t] [-r rand_seed] [ -p read_profile ] [ -c num_flow_cycles ] <INPUT_SEQ_FILE>
<OUTPUT_FILE_PREFIX> <FOLD_COVERAGE> <MEAN_FRAG_LEN> <STD_DEV>

AMPLICON SEQUENCING SIMULATION
        art_454 [-s] [-a ] [-t] [-r rand_seed] [ -p read_profile ] [ -c num_flow_cycles ] <-A|-B> <INPUT_SE
Q_FILE> <OUTPUT_FILE_PREFIX> <#_READS/#_READ_PAIRS_PER_AMPLICON>

===== PARAMETERS =====

INPUT_SEQ_FILE            -  the filename of DNA/RNA reference sequences in FASTA format
OUTPUT_FILE_PREFIX        -  the prefix or directory of output read data file (*.fq) and read alignment file
 (*.aln)
FOLD_COVERAGE             -  the fold of read coverage over the reference sequences
MEAN_FRAG_LEN             -  the average DNA fragment size for paired-end read simulation
STD_DEV                   -  the standard deviation of the DNA fragment size for paired-end read simulation
#READS_PER_AMPLICON       -  number of reads per amplicon (for 5'end amplicon sequencing)
#READ_PAIRS_PER_AMPLICON  -  number of read pairs per amplicon (for two-end amplicon sequencing)

===== OPTIONAL PARAMETERS =====

-A indicate to perform single-end amplicon sequencing simulation
-B indicate to perform paired-end amplicon sequencing simulation
-M indicate to use CIGAR 'M' instead of '=/X' for alignment match/mismatch
-a indicate to output the ALN alignment file
-s indicate to output the SAM alignment file
-d print out warning messages for debugging
-t indicate to simulate reads from the built-in GS FLX Titanium profile [default: GS FLX profile]
-r specify a fixed random seed for the simulation (to generate two identical datasets from two different ru
ns)
-c specify the number of flow cycles by the sequencer [ default: 100 for GS-FLX, and 200 for GS-FLX Titaniu
m ]
-p specify user's own read profile for simulation
   NOTE: the name of a read profile is the directory containing read profile data files.
         please read the REAME file about the format of 454 read profile data files and.
         and the default filenames of these data files.
```

Figure 8. Art_454 tool.

The tool's parameters allow the simulation of *single-end*, *paired-end* and *amplicon* sequencing runs. It is possible to determine coverage, fragment size, standard deviation for fragment size, number of *reads* per *amplicon* and number of *read* pairs per *amplicon*.

The art_SOLiD tool allows the simulation of *single-end, matepair, paired-end* and *amplicon* sequencing runs (Figure 9). The following parameters are required:

- INPUT_SEQ_FILE: file in FASTA format containing the DNA or RNA reference sequences;
- OUTPUT_FILE_PREFIX: the prefix of the files generated by the simulator;
- FOLD_COVERAGE: the sequencing coverage of the reference genome;
- LEN_READ: size of F3/R3 *reads;*
- LEN_READ_F3: size of F3 *reads* for *paired-end* race simulations;
- LEN_READ_F5: size of F5 *reads* for *paired-end* race simulations;
- MEAN_FRAG_LEN: the average DNA fragment size for *paired-end* and *matepair* race simulations*;*
- STD_DEV: the standard deviation for DNA fragment size for *paired-end* and *matepair* race simulations*;*
- #READS_PER_AMPLICON: number of *reads* per *amplicon* ;
- #READ_PAIRS_PER_AMPLICON: number of *read* pairs per *amplicon* ;

The art_illumina tool allows the simulation of *single-end*, *matepair*, *paired-end* and *amplicon* sequencing runs (Figure 10). The following parameters are required:

- i: file in FASTA format containing the DNA or RNA reference sequences;
- The prefix of the files generated by the simulator;
- f:the sequencing coverage of the reference genome;
- l: size of the *reads* to be simulated;
- m: the average DNA fragment size for *paired-end* race simulations*;*
- s: the standard deviation for DNA fragment size for *paired-end* race simulations*;*

- sam: generates a SAM alignment file;

```
====================================================================
                   ART_SOLiD (Version 1.3.3)
        Simulation of Applied Biosystems' SOLiD Sequencing
   Copyright (c) 2008-2015, Weichun Huang. All Rights Reserved.
====================================================================

===== USAGES =====

SINGLE-END (F3 READ) SIMULATION
        art_SOLiD [ options ] <INPUT_SEQ_FILE> <OUTPUT_FILE_PREFIX> <LEN_READ> <FOLD_COVERAGE>

MATE-PAIR READS (F3-R3 PAIR) SIMULATION
        art_SOLiD [ options ] <INPUT_SEQ_FILE> <OUTPUT_FILE_PREFIX> <LEN_READ> <FOLD_COVERAGE> <MEAN_FRAG_L
EN> <STD_DEV>

PAIRED-END READS (F3-F5 PAIR) SIMULATION
        art_SOLiD [ options ] <INPUT_SEQ_FILE> <OUTPUT_FILE_PREFIX> <LEN_READ_F3> <LEN_READ_F5> <FOLD_COVER
AGE> <MEAN_FRAG_LEN> <STD_DEV>

AMPLICON SEQUENCING SIMULATION
        art_SOLiD [ options ] -A s <INPUT_SEQ_FILE> <OUTPUT_FILE_PREFIX> <LEN_READ> <READS_PER_AMPLICON>
        art_SOLiD [ options ] -A m <INPUT_SEQ_FILE> <OUTPUT_FILE_PREFIX> <LEN_READ> <READ_PAIRS_PER_AMPLICO
N>
        art_SOLiD [ options ] -A p <INPUT_SEQ_FILE> <OUTPUT_FILE_PREFIX> <LEN_READ_F3> <LEN_READ_F5> <READ_
PAIRS_PER_AMPLICON>

===== PARAMETERS =====

INPUT_SEQ_FILE           -  filename of DNA/RNA reference sequences in FASTA format
OUTPUT_FILE_PREFIX       -  prefix or directory for all output read data files
FOLD_COVERAGE            -  fold of read coverage over the reference sequences
LEN_READ                 -  length of F3/R3 reads
LEN_READ_F3              -  length of F3 reads for paired-end read simulation
LEN_READ_F5              -  length of F5 reads for paired-end read simulation
READS_PER_AMPLICON       -  number of reads per amplicon
READ_PAIRS_PER_AMPLICON  -  number of read pairs per amplicon
MEAN_FRAG_LEN            -  mean DNA/RNA fragment size for matepair/paired-end read simulation
STD_DEV                  -  standard deviation of the DNA/RNA fragment sizes for matepair/paired-end read
simulation

===== OPTIONAL PARAMETERS =====

-A specify the read type for amplicon sequencing simulation (s:single-end, m: matepair, p: paired-end)
-M indicate to use CIGAR 'M' instead of '=/X' for alignment match/mismatch
-s indicate to generate a SAM alignment file
-r specify the random seed for the simulation
-f specify the scale factor adjusting error rate (e.g., -f 0 for zero-error rate simulation)
-p specify user's own read profile for simulation
```

Figure 9. art_SOLiD tool.

The tool's parameters allow the simulation of *single-end*, *paired-end*, *matepair* and *amplicon* sequencing runs. It is possible to determine coverage, F3/R3, F3 and F5 *read* sizes, fragment size, standard deviation for fragment size, number of *reads* per *amplicon* and number of *read* pairs per *amplicon*.

```
====================================ART====================================
                          ART_Illumina (2008-2015)
                          Q Version 2.3.7 (Mar 19, 2015)
                    Contact: Weichun Huang at whduke@gmail.com
===========================================================================

===== USAGE =====

art_illumina [options] -sam -i <seq_ref_file> -l <read_length> -f <fold_coverage> -ss <sequencing_system> -
o <outfile_prefix>
art_illumina [options] -sam -i <seq_ref_file> -l <read_length> -f <fold_coverage> -o <outfile_prefix>
art_illumina [options] -sam -i <seq_ref_file> -l <read_length> -c <total_num_reads> -o <outfile_prefix>
art_illumina [options] -sam -i <seq_ref_file> -l <read_length> -f <fold_coverage> -m <mean_fragsize> -s <st
d_fragsize> -o <outfile_prefix>
art_illumina [options] -sam -i <seq_ref_file> -l <read_length> -c <total_num_reads> -m <mean_fragsize> -s <
std_fragsize> -o <outfile_prefix>

===== PARAMETERS =====

  -1   --qprof1   the first-read quality profile
  -2   --qprof2   the second-read quality profile
  -amp --amplicon amplicon sequencing simulation
  -c   --rcount   total number of reads/read pairs to be generated [per amplicon if for amplicon simulation
](not be used together with -f/--fcov)
  -d   --id       the prefix identification tag for read ID
  -ef  --errfree  indicate to generate the zero sequencing errors SAM file as well the regular one
                  NOTE: the reads in the zero-error SAM file have the same alignment positions
                  as those in the regular SAM file, but have no sequencing errors
  -f   --fcov     the fold of read coverage to be simulated or number of reads/read pairs generated for eac
h amplicon
  -h   --help     print out usage information
  -i   --in       the filename of input DNA/RNA reference
  -ir  --insRate  the first-read insertion rate (default: 0.00009)
  -ir2 --insRate2 the second-read insertion rate (default: 0.00015)
  -dr  --delRate  the first-read deletion rate (default:  0.00011)
  -dr2 --delRate2 the second-read deletion rate (default: 0.00023)
  -l   --len      the length of reads to be simulated
  -m   --mflen    the mean size of DNA/RNA fragments for paired-end simulations
  -mp  --matepair indicate a mate-pair read simulation
  -nf  --maskN    the cutoff frequency of 'N' in a window size of the read length for masking genomic regio
ns
                  NOTE: default: '-nf 1' to mask all regions with 'N'. Use '-nf 0' to turn off masking
  -na  --noALN    do not output ALN alignment file
  -o   --out      the prefix of output filename
  -p   --paired   indicate a paired-end read simulation or to generate reads from both ends of amplicons
                  NOTE: art will automatically switch to a mate-pair simulation if the given mean fragment
size >= 2000

  -q   --quiet    turn off end of run summary
  -qs  --qShift   the amount to shift every first-read quality score by
  -qs2 --qShift2  the amount to shift every second-read quality score by
                  NOTE: For -qs/-qs2 option, a positive number will shift up quality scores (the max is 93)

                  that reduce substitution sequencing errors and a negative number will shift down
                  quality scores that increase sequencing errors. If shifting scores by x, the error
                  rate will be 1/(10^(x/10)) of the default profile.
  -rs  --rndSeed  the seed for random number generator (default: system time in second)
                  NOTE: using a fixed seed to generate two identical datasets from different runs
  -s   --sdev     the standard deviation of DNA/RNA fragment size for paired-end simulations.
  -sam --samout   indicate to generate SAM alignment file
  -sp  --sepProf  indicate to use separate quality profiles for different bases (ATGC)
  -ss  --seqSys   The name of Illumina sequencing system of the built-in profile used for simulation
                  NOTE: sequencing system id names are:
                        GA1 - Genome Analyzer I, GA2 - Genome Analyzer II
                        HS10 - HiSeq 1000, HS20 - HiSeq 2000, HS25 - HiSeq 2500, MS - MiSeq
  -M   --cigarM   indicate to use CIGAR 'M' instead of '=/X' for alignment match/mismatch
```

Figure 10. Artjllumina tool.

The tool's parameters allow the simulation of *single-end*, *paired-end*, *matepair* and *amplicon* sequencing runs. It is possible to determine coverage, *read* sizes, fragment sizes, standard deviation for fragment sizes and whether an alignment file should be generated.

The *BEAR Simulator* uses a machine learning approach to generate leituras with a size and quality close to empirically derived values. The approach used by the *BEAR* simulator to generate data eliminates the need to refine the parameters, allowing for an easy-to-use interface, so that the user only needs to enter a sample of data that has the desired properties to generate the emulated readings (Johnson *et al*, 2014). The BEAR simulator generates data from the Illumina, 454 and Ion Torrent sequencing platforms.

Simulation in BEAR takes place using a set of *scripts* developed in Perl and Python (version 1.0). These *scripts* implement the 4 phases required to generate an NGS run in this simulator: generation of the homology-based profile, generation of the error model, generation of the quality *score* model and generation of the simulation readings.

In the work of (Schneeberger, 2014), a list of representative works of the *mapping-by-sequencing* methodology is presented. These papers use one of four sequencing platforms in their experiments: 454, SOLID, *Illumina* and *Ion Torrent*. As there is no simulator that generates data from these four platforms, it is necessary to combine two of them, in this case the *ART* (for 454, SOLID and *Illumina* data) and *BEAR* (for *Ion Torrent* data) simulators.

3.2 BOWTIE AND SAMTOOLS

The alignment and mapping (Figure 6-4) of NGS sequencing *reads* to a reference genome is the first analysis step many comparative genomics *pipelines*. In many cases, this step is the slowest, because for each read the alignment tool needs to solve a complex computational problem: determining the probability of the start point of the read according to the reference genome (Langmead & Salzberg, 2012). There are several specialized tools for this stage, some of which are more specialized, such as Stampy (Lunter & Goodson, 2010), specific for Illumina, SNP-o-matic (Manske & Kwiatkowski, 2009), which emphasizes genotyping, and others with a general purpose such as NextGenMap (Sedlazeck, Rescheneder P & von Haeseler, 2013), BWA (Li & Durbin, 2009) and Bowtie (Langmead & Salzberg, 2012), the latter being the choice for this work.

Bowtie has more than 30 parameters that can be used to map *reads* the reference genome. Of these parameters, only four are mandatory: prefix of the index files (-x), files with the sequencing *reads* (-1 and -2 - for *paired-end* runs) and the name of the SAM file that will be generated. This SAM file contains the alignment data.

Once the alignment has been made, some conversions need to take place until the NGM-specific file is reached (Figure 6-5). These conversions are made by the tools in the SAMtools suite. This suite of tools offers utilities for manipulating SAM files, including sorting, merging and indexing (Li *et al*, 2009).

First, the SAM file is converted into an ordered BAM file (binary version of the SAM file). This conversion makes it easier to manipulate the alignment data. Then, *base calling* is done with the *pileup* tool from the SAMtools suite. The *pileup* format describes the base pair information at each chromosome position and is used to identify SNPs. Finally, the *pileup* file is converted to EMAP format using a script developed in the Perl language, SAM2NGM.pl. This script was developed and distributed as part of the use of NGM. After conversion, the final EMAP file is inserted into the NGM tool for analysis and identification of SNPs.

3.3 NEXT-GENERATION MAPPING

In 2011, Austin and colleagues developed a new *mapping-by-sequencing* methodology. This methodology, called *Next-generation Mapping* (NGM), requires no prior mapping knowledge and only a small number of F2 populations. In this work, three new genes related to the composition of the *Arabidopsis thaliana* cell wall were identified (Austin *et al*, 2009). This technique was used in the work of Allen *et al.* (Allen *et al*, 2013) and Tabata *et al.* (Tabata *et al*, 2012), who successfully identified mutations using 110 and 12 F2 mutant lines, respectively.

The *web* tool developed allows fast and robust mapping of EMS mutations via NGM. This tool works in four stages (Austin, 2013):

- pre-processing, sending and filtering the SNP data obtained from the sequenced F2 mapping population;
- locating the mutation on a specific chromosome by identifying the non-recombinant region;
- SNP frequency partitioning based on degrees of discordance with the reference genome;

- localization of the causal mutation and annotation of candidate SNPs within a mapped region.

Each stage allows for user intervention and dynamic adjustments with the possibility of returning to previous stages. The algorithms used for mutation mapping are implemented in the R statistical programming language.

NGM can be applied to any organism with a well-studied genome that can be crossed to generate a mapping strain. Furthermore, this technique can be applied to a small population, which makes it easier and cheaper to use in the laboratory (Austin *et al*, 2011). However, one point to note is the DNA *pooled* sequencing platform. The sequencing platform and parameters have always been fixed in the works studied.

3.4 COMPUTER RESOURCES

The WEB tool was developed using the Linux platform and the Apache 2 web server. The graphical user interface was created by programming in PHP, MySQL, HTML and *JavaScript*.

4. RESULTS AND DISCUSSION

4.1 NGMDATASIM *PIPELINE*

Based on the architecture analyzed, it is possible to see that *mapping-by-sequencing* experiments can vary according to the design of the experiment. However, it is possible to observe a pattern in these experiments. *In* this way, an *in-silico* bioinformatics *pipeline* is proposed, with the aim of providing research, analysis and optimization of parameters (size of the mapping population, concentration of EMS to be used, sequencing coverage required, size of the sequencing reads required, best sequencing platform for the experiment, use of *paired-end* or *single-end* reads in the sequencing, etc.) and possibly generating better project designs.) and possibly generate better designs for *mapping-by-sequencing* projects that use NGM in *Arabidopsis thaliana*. This *pipeline* is divided into 6 parts (Figure 6) and has been discussed previously.

To make it easier to use this *pipeline*, a graphical interface was developed which can be accessed via the internet from any computer, *tablet* or *smartphone*. The implementation of the 6 phases of the *pipeline* were grouped into 4 steps in the development of the graphic interface: creation of the population, mutation parameters, NGS sequencing and choice of output type.

In the first stage, creating the mapping population, the title of the experiment, the number of plants and the number of mutant plants in the mapping population must be defined, as well as the crossing scheme that will give rise to this population (*outcross* or *backcross*) (Figure 11).

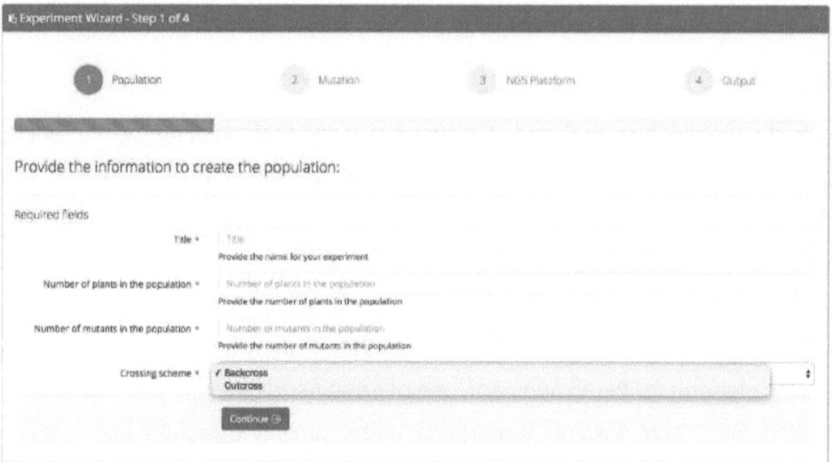

Figure 11. First stage of NGMdatasim: creation of the mapping population.

Definition of the experiment title, number of plants and number of mutant plants in the mapping population and the crossing scheme.

In the second step, mutation parameters, the mutation density to be generated, the target mutation, the reference genome base as well as the mutant base for the target mutation must be defined (Figure 12).

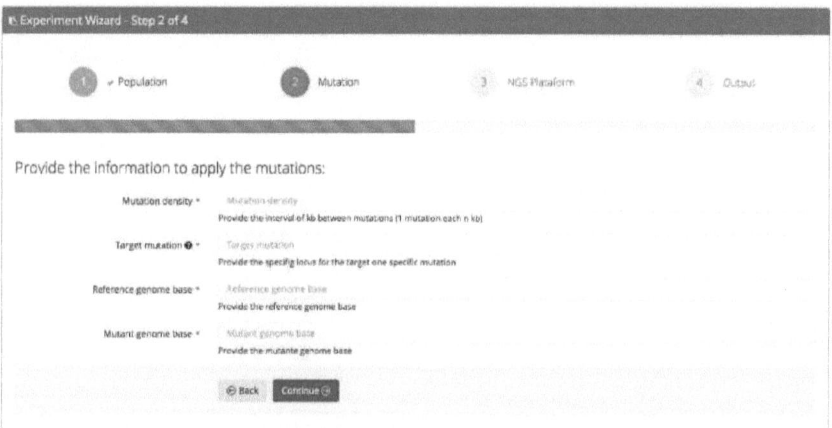

Figure 12. Second stage of NGMdatasim: mutation parameters.

Definition of the mutation density, the target mutation, the reference base and the mutant base to be generated.

In the third step, NGS Sequencing, a sequencing platform must be chosen, as well as the extra parameters for each platform (Figure 13). The platforms defined for simulating NGS runs are: 454 Roche, SOLiD, *Illumina's Genome Analyzer* I and II, *Illumina's HiSeq* 1000, 2000, 2500, *MiSeq* and *Ion Torrent* PGM. The parameters that must be set for each platform have been specified in section 1.4.2.

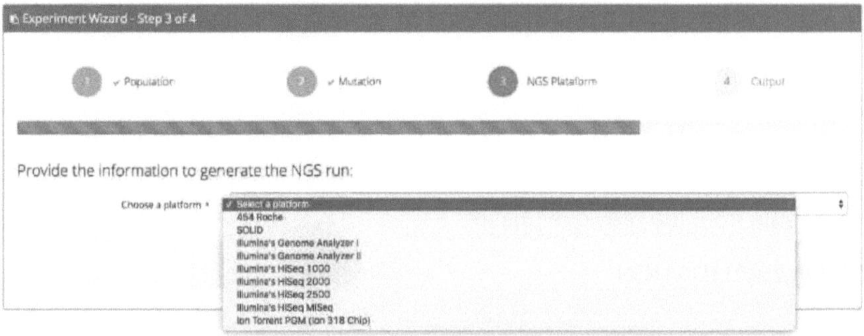

Figure 13. Third stage of NGMdatasim: NGS sequencing.

Definition of the NGS platform to be used and the parameters specific to the selected platform.

In the last step, choosing the type of output, it must be defined whether the output from the processing of the previous steps should be the format for NGM processing or the raw processing data (Figure 14). The output file for NGM has the extension EMAP and summarizes all the data from the previous processing. The raw data comprises: the run files (in FASTAQ format), the alignment file between the *reads* and the reference genome (in SAM format), the sorted BAM file, the *base calling* file (in PILEUP format) and the NGM processing file (in EMAP format). These files can be downloaded and manipulated separately by the user.

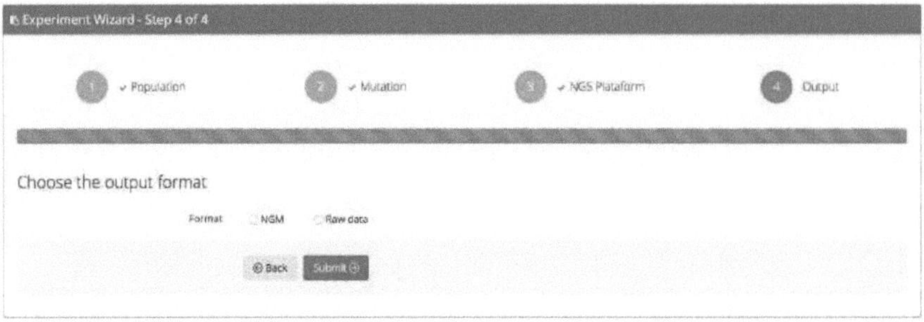

Figure 14. Fourth stage of NGMdatasim: , choice of output type.

Define whether the *pipeline*'s processing output should be the NGM file or the raw processing data from the previous phases.

4.2 *PIPELINE* VALIDATION

In order to assess the degree of reliability of the data generated by the *pipeline*, a validation of the *pipeline* data was developed, consisting of comparing the data generated for three mutations defined in the work by (Austin *et al*, 2011). These three mutations are related in the cell wall of *Arabidopsis thaliana* and are located in three different genes: *mur11*, which is located on chromosome 3 at position 22084518 and makes a C to T base substitution, *fphl*, which is located on chromosome 1 at position 30206068 and makes a G to A base substitution and *fph2*, which is located on chromosome 1 at position 22814777 and makes a C to T base substitution.

In the work by Austin and colleagues, three mapping populations are defined, one for each mutation, containing 80 mutants each, formed from *backcrosses* sequenced on *Illumina's Genome Analyzer* platform, with 30x coverage and average *read* sizes of 38 bp. A simulation was then carried out with this data, i.e. for each mutation described, a mapping population formed by *backcrosses* was generated, the mutations applied, the genomes sequenced and the results analyzed. In addition to the mutation of interest, another 2,500 random mutations were applied to all the genomes.

In addition, two other populations, containing 50 and 10 mutants each, were generated for each mutation. The idea is to evaluate the effectiveness of the NGM tool in even smaller populations. This was done because, in the work by Austin et al., a minimum number of 50 individuals is indicated for forming the mapping population, but the tool could work, in theory, on an even smaller number, with 10 mutants, for example. In addition, a mutant was generated for direct sequencing and analysis of the results. Table 3 shows the data observed in the experiments as a whole, the results of the NGM experiments by Austin et al. are shown in Table 4 and the comparison of results is shown in Table 5.

Table 3. Summary of data from *pipeline* validation experiments.

Change	Position	Replacement	Population	*Reads*	Mapped *reads*
mur11 *	3:22084518	C-T	80	115610496	75111880
mur11	3:22084518	C-T	80	232079378	212878207
mur11	3:22084518	C-T	50	148675838	137073812
mur11	3:22084518	C-T	10	91493828	84802581
mur11	3:22084518	C-T	1	93899861	87481531
fph1 *	1:30206068	G-A	80	302978858	215745005
fph1	1:30206068	G-A	80	169911492	160104837
fph1	1:30206068	G-A	50	116606690	108094194
fph1	1:30206068	G-A	10	74337890	67667653
fph1	1:30206068	G-A	1	75918532	69223523
fph2 *	1:22814777	C-T	80	148509664	122778391
fph2	1:22814777	C-T	80	142013990	130723104
fph2	1:22814777	C-T	50	90103326	83096158
fph2	1:22814777	C-T	10	85493976	81465642
fph2	1:22814777	C-T	1	81732380	79262675

Legend: (*) Data from experiments by (Austin *et al*, 2011).

Table 4: Mutations found in the experiments of (Austin *et al,* 2011).

Change	Position	Change
mur11	3:20795011	A-T
mur11	3:20795012	T-C
mur11	3:22003267	C-T
mur11 *	3:22084518	C-T
mur11	3:22752769	G-A
fph1 *	1:30206068	G-A
fph2 *	1:22814777	C-T
fph2	1:22926537	C-T

Caption: (*) mutations verified in a bench experiment. Adapted from (Austin *et al,* 2011).

Table 5: Mutations found *in the in vivo* experiments compared to the mutations found *in the in vitro* experiments.

Change	Real data	Simulated Data			
	80 mutants	1 mutant	10 mutants	50 mutants	80 mutants
	N. SNPs	N. SNPs	N. SNPs	N. SNPs	N. SNPs
mur 11	5	0 *	11	6	9
fph1	1	0 *	3	5	11
fph2	2	0 *	42	5	9

Legend: (*) unidentified mutations.

Based on the data generated with the NGMdatasim$_1$ *pipeline*, it was possible to find all the mutations described in the work by Austin et al. in all the mapping populations, with the exception of the direct sequencing of the mutant. It is assumed that the detection of SNPs from just one mutant is not possible because the tool uses statistical inferences about allele frequency to detect SNPs and these inferences cannot be made in just one mutant.

In the population of 10 mutants, the results were discrepant for all populations; from 3 to 42 mutations were found. It is assumed that the same argument for the non-detection of SNPs in the direct sequencing of mutants can be used. The most consistent results for the detection of SNPs were found in the 50 and 80 mutant populations. It is interesting to note that the number of SNPs found in the population of 50 mutants was lower than the number of SNPs found in the population of 80 mutants. This could be an interesting indication to be analyzed *in an in vivo* experiment.

In general, the results of the data simulations were satisfactory and achieved the objective they set out to achieve. A total of 611 gigabytes of data were produced in the 12 comparative experiments presented, which took 2 months to process. It is clear that the use of data simulation can offer good indications for decision-making in biological experiments, since the time and cost of *in vivo* experiments are much lower than *in vitro* experiments.

5. FINAL CONSIDERATIONS

In this work, a bioinformatics *pipeline* was defined and implemented for simulating data from NGM experiments in *Arabidopsis thaliana*. This pipeline consists of four phases: emulation of the mapping population, simulation of mutants in the mapping population, NGS runs of the mutant genomes and mapping between the NGS reads and the *Arabidopsis thaliana* reference genome. To facilitate access and use, a graphical interface has been developed for the *pipeline* that can be accessed via the Internet.

The *pipeline* developed generates a specific file for processing in the NGM tool, which identifies the point mutation referring to the mutant phenotype of the mapping population.

The *pipeline* was validated by comparing data from Austin's experiment (2011) with data simulated by the *pipeline*. It was possible to verify that the use of a smaller number of mutants (50) compared to the number of the population in Austin's work (2011) (80) generates better results.

The results of the *pipeline* tests were satisfactory. However, there is room for improvement in the simulation scenarios. Every simulation is based on a set of variables observable in real experiments. It would therefore be important for these variables to be clearly defined in the context of *mapping-by-sequencing* and NGM experiments so that data simulations can be increasingly accurate. In addition, the use of artificial intelligence and machine learning techniques could make the results even more consistent and reliable.

6. REFERENCES

AbeA$_1$ Kosugi S, Yoshida K, et al. (2011) Genome sequencing reveals agronomically important loci in rice using MutMap. 2011, Nature Biotechnology, Volume 30, 174-178.

Allen RS, Nakasugi K, Doran RL, *et al.* (2013) Facile mutant identification via a single parental backcross method and application of whole genome sequencing based mapping pipelines. 2013, Frontiers in plant science, Volume 4.

Arabidopsis Genome Initiative (2000). Nature 408, 796-815 (December 14, 2000)

Austin RS, Vidaurre D, Stamatiou G, et al. (2011) Next-generation mapping of Arabidopsis genes. The PlantJournal. 2011, Volume 67, 715-725.

Bowen ME, Henke K, Siegfried KR, Warman ML, *et al.* (2012) Efficient mapping and cloning of mutations in zebrafish by low-coverage whole-genome sequencing. 2012, Genetics, Vol. 190, 1017-1024.

Bragg LM, Stone G, Butler MK, et al. (2013). Shining a Light on Dark Sequencing: Characterizing Errors in Ion Torrent PGM data. PLoS Comp. Biol., 2013.

Brodin J, Mild M, Hedskog C, et al. (2013). PCR-Induced Transitions Are the Major Source of Error in Cleaned Ultra-Deep Pyrosequencing Data. PLoS One, 2013.

Chen YC, Liu T, Yu CH, et al. (2013). Effects of GC Bias in Next-GenerationSequencing Data on De Novo Genome Assembly. PLoS One, 8(4):e62856, 2013.

Dabney J, Meyer M. (2012). Length and GC-biases during sequencing library amplification: A comparison of various polymerase-buffer systems with ancient and modern DNA sequencing libraries. BioTechniques, 52(2):87-94, 2012.

Edwards MD, Gifford DK. (2012) High-resolution genetic mapping with pooled sequencing. 2012, BMC Bioinformatics, Volume 13.

Fekih R, Takagi H, Tamiru M, et al. (2013) MutMap+: Genetic Mapping and Mutant Identification without Crossing in Rice. 2013, PLOS ONE, Volume 8, Issue 7.

Flusberg BA, Webster DR, Lee JH, et al. (2010) Direct detection of DNA methylation during single-molecule, real-time sequencing. 2010, Nature Methods 7, 461 465.

Gabriel C., Furst D., Fae I., et al. (2014). Hla typing by nextgeneration sequencing getting closer to reality. Tissue Antigens, 83(2):65-75, 2014.

Glenn T. (2011) Field guide to next-generation DNA sequencers. 2011, Molecular ecology resources. Volume 11, Issue 5.

Goias, TM., Geest H., Gros, J., et al. (2012). Comparative next-generation mapping of the Phytophthora infestans resistance gene Rpi-dlc2 in a European accession ofSolanum dulcamara. TheorAppl Genet. 20l3.

Guo Y, Li C, Long C, et al. (2012). The effect of strand bias in illumina short-read sequencing data. BMC Genomics, 13:666, 2012.

Hartwig B, James GV, Konrad K, et al. (2012) Fast Isogenic Mapping-by-Sequencing of Ethyl Methanesulfonate-Induced Mutant Bulks. Plant Physiology. 2012, Volume 160, 591-600.

Huang W, Li L, Myers RJ, et al. (2011). ART: a next-generation sequencing read simulator. 2011, BIOINFORMATICS APPLICATIONS NOTE, Vol. 28 no. 4 2012, pages 593-594.

International Human Genome Consortium. (2004) Finishing the euchromatic sequence of the human genome. 2004, Nature 431, 931-945.

James GV, Patel V, Nordstrom KJV, et al. (2013) User guide for mapping-by-sequencing in Arabidopsis. Genome Biology 2013, 14:R61.

Jander G, Norris SR, Rounsley SD, et al. (2002) Arabidopsis Map-Based Cloning in the Post-Genome Era. Plant Physiol. 2002, Volume 129.

Johnson S, Trost B, Long RJ, et al. (2014) A better sequence-read simulator program for metagenomics. 2014, BMC Bioinformatics 2014 15(Suppl 9):S14.

Koornneef M, Meinke D. (2010) The development of Arabidopsis as a model plant. 2010, The PlantJournal 61, 909-921.

Langmead B, Salzberg SL. (2012) Fast gapped-read alignment with Bowtie 2. Nat Methods. 2012 Mar4.

Leshchiner I, Alexa K, Kelsey P, et al. (2012) Mutation mapping and identification by whole-genome sequencing. 2012, Genome Research ,Volume 22, 1541-1548.

Li H, Durbin R. (2009) Fast and accurate short read alignment with Burrows-Wheeler transform. Bioinformatics.

Li H, Handsaker B, Wysoker A, et al. (2009) The Sequence alignment/map (SAM) format and SAMtools. Bioinformatics, 25, 2078-9.

Lindner H, Raissig MT, Sailer C, et al. (2012) SNP-Ratio Mapping (SRM): identifying lethal alleles and mutations in complex genetic backgrounds by next-generation sequencing. 2012, Genetics, Volume 191, Issue 4.

Lightfield J, Fram NR, Ely B. (2011). Across Bacterial Phyla, Distantly-Related Genomes with Similar Genomic GC Content Have Similar Patterns of Amino Acid Usage. PLoS One, 6(3):e17677, 2011.

Liu L, Li Y, Li S, et al. (2012). Comparison of next-generation sequencing systems. 2012, Journal of biomedicine & biotechnology, Volume 2012, Article ID 251364.

Lunter G, Goodson M. (2011) Stampy: a statistical algorithm for sensitive and fast mapping of Illumina sequence reads. Genome Res. 2011 Jun.

Manske HM1, Kwiatkowski DP. (2009) SNP-o-matic. Bioinformatics.

Mardis ER. (2008). The impact of next-generation sequencing technology on genetics. 2008, Trends in Genetics, Volume 24.

McCallum, CM., Comai, L., Greene, et al. (2000) Targeted screening for induced mutations. Nat. Biotech. 18, 455-457.

Meinke, DW. (2013). Trends in Plant Science, 2013. Volume 18, Issue 2, p84-91

Metzker ML. (2010) Sequencing technologies - the next generation. Nature Reviews Genetics, 2010, Volume 11.

Michelmore RW, Param I, Kesseli RV. (1991) Identification of markers linked to disease-resistance genes by bulked segregant analysis: a rapid method to detect markers in specific genomic regions by using segregating populations. Proceedings of the National Academy of Sciences of the United States of America. 1991, Volume 88. No 21.

Minevich G, Park DS, Blankenberg D, et al. (2012) CloudMap: a cloud-based pipeline for analysis of mutant genome sequences. 2012, Genetics, Volume 192, 12491269.

Obholzer N, Swinburne IA, Schwab E, et al. (2012) Rapid positional cloning of zebrafish mutations by linkage and homozygosity mapping using whole-genome sequencing. 2012, Development, Volume 139, 4280-4290.

Page DR, Grossniklaus U. (2002) The art and design of genetic screens: Arabidopsis thaliana. Nature Reviews Genetics. 2002, Volume 3.

Pettersson E., Lundeberg J., Ahmadian A. (2009) Generations of sequencing technologies. Genomics 93(2): 105-111

Quail MA, Smith M, Coupland P, et al. (2012). A tale of three next generation sequencing platforms: comparison of Ion Torrent, Pacific Biosciences and Illumina MiSeq sequencers. BMC Genomics, 13(1):341, 2012.

Rafalski JA. (2010)Association genetics in crop improvement.Current Opinion in Plant Biology. 2010, 13: 174-180.

Ronaghi M., Uhlen M., Nyren P. (1998). A Sequencing Method Based on Real-Time Pyrophosphate. Science, 281(5375):363-365, 1998.

Rusk N. Torrents of sequence. Nature Methods, 8(44):doi:10.1038/nmeth.f.330, 2011.

Sanger F, Nicklen S, Coulson AR. 1977. DNA sequencing with chain-terminating inhibitors. Proceedings of the National Academy of Sciences, U SA74: 5463-7.

Schirmer M, Damore L, Hall N, et al. (2013). Error profiles for next generation sequencing technologies. EMBnetjournal, 19:81-83, 2013.

Schneeberger K, Ossowski S, Lanz C, et al. (2009), SHOREmap: simultaneous mapping and mutation identification by deep sequencing. Nature Methods. 2009, Volume 6, No. 08.

Schneeberger K, Weigel D. (2011) Fast-forward genetics enabled by new sequencing technologies. Trend in Plant Science. 2011, Volume. 16, No. 05.

Schneeberger K. (2014). Using next-generation sequencing to isolate mutant genes from forward genetic screens. Nature Reviews Genetics. 2014, Volume 15.

Sedlazeck FJ, Rescheneder P, von Haeseler A. (2013) NextGenMap: fast and accurate read mapping in highly polymorphic genomes. Bioinformatics.

Sun X, Peng T, Mumm RH. (2011). The role and basics of computer simulation in support ofcritical decisions in plant breeding. Mol Breeding (2011).

Tabata R, Kamiya T, Shigenobu S, et al. (2012) Identification of an EMS-induced causal mutation in a gene required for boron-mediated root development by low-coverage genome re-sequencing in Arabidopsis. 2012, Plant signaling & behavior, Volume 8, Issue 1.

Verbruggen N, Hanikenne M, Clemens S. (2013). A more complete picture of metal hyperaccumulation through next-generation sequencing technologies. 2013, Frontiers in plant science, Volume 4.

MIX
Papier aus verantwortungsvollen Quellen
Paper from responsible sources
FSC® C105338

Printed by Books on Demand GmbH, Norderstedt / Germany